Practical Instrumental Analysis

Dr. MANISH KUMAR THIMMARAJU

M.Pharm, Ph.D

Edited By

Mr. KHAGGESWAR BHEEMANAPALLY

M.Pharm (BITS-Pilani)

ISBN-13: 978-1535094214

ISBN-10: 1535094214

DEDICATION

To my parents, teachers and students

DISCLAIMER

The authors, persons, organizations or anything mentioned in this textbook does not take any liability for accidents, damages, profit or loss arise out of following mentioned experiments and or literature.

ABOUT AUTHOR

Dr.Manish Kumar Thimmaraju pursued Master in Pharmacy

(2004-2006), Pharmaceutical Analysis as Specialization from Dr.MGR Medical University, Chennai. He obtained PhD in 2014 from Acharya Nagarjuna University, Andhra Pradesh, India. He has published 38 scientific research papers in reputed national and international journals and presented more than 20 abstracts in national and international scientific conferences in the area of Pharmaceutical Analysis. He is authored a book titled "Handbook of UV-Visible and HPLC for Beginners" and to his credit he has filed 03 national patents. He is a life member of Association of pharmaceutical teachers of India, Member in Association of Pharmacy and Biotechnology and Society of biotechnologists, India. At present Dr. Manish Kumar Thimmaraju is serving as Head of the department, Pharmaceutical Analysis, Balaji Institute of Pharmaceutical Sciences, Warangal, Telangana, India.

ACKNOWLEDGMENTS

I express my thanks to **Mr. Pavan Kumar Thimmaraju,** who was responsible for proof reading of the manuscript, I owe great indebtness.

The initiation and completion of this work was speeded up due to the deep concern and regular monitoring done by **Dr. J. Venkateswar Rao**, Principal, Talla Padmavathi College of Pharmacy, Warangal.

-Author

Contents

1. RADIAL AND ASCENDING PAPER CHROMATOGRAPHY

Dr. Manish Kumar Thimmaraju

Aim:

To perform and submit radial and ascending paper chromatography and identify the unknown components present in the given sample.

Apparatus required:

Whatmann filter paper, glass beakers, petriplates, 10 ml measuring cylinder and capillary tubes, etc.

Chemicals required:

n-butanol, glacial acetic acid, distilled water (4:1:5), amino acids (valine and proline) and ninhydrin reagent.

Principle:

Chromatography is a technique used to isolate individual components from complex mixtures. This method involves the interaction of components between stationary phase and mobile phase. In this technique, stationary phase is paper and mobile phase is usually a solvent which drives the mixture of molecules on the paper towards end of the paper. Partition plays an important role in the paper chromatography than adsorption.

Retention factor (Rf):

It is defined as the ration of distance travelled by the solute (single drug spot) from baseline to the distance travelled by the mobile phase. R_f value lies between zero and one, and also never exceeds one and has no units.

R_f = distance travelled by drug substance / distance travelled by mobile phase

1

Reaction mechanism:

Ninhydrin and amino acids react with each other to produce purple or yellow color.

| Ninhydrin Reagent | Amino acid | | Purple complex |

In a series of reactions

The types of paper chromatography are

1. Radial paper chromatography
2. Descending paper chromatography
3. Ascending paper chromatography
4. Ascending and descending chromatography
5. Two dimensional paper chromatography

Advantages of paper chromatography:

1. Paper chromatography is the simplest of all methods.
2. It is very inexpensive to perform in the lab.
3. Very little sample is required.

Disadvantages of paper chromatography:

1. The method is not applicable for quantitative analysis.
2. This method is not suitable for resolution(separation) of complex mixtures.

Applications:

Paper chromatography is used for the separation of

➢ polar and non polar compounds.

2

- amino acids, carbohydrates, vitamins, antibiotics, alkaloids, glycosides etc.
- organic compounds, bio-chemicals.
- identification of hormones, drugs.
- Identification of impurities.
- Identification of unknown substances in drugs
- Identification of decomposition products.

Solvent preparation:

1. Saturate n-butanol with water for 24 hours.
2. Separate above using funnel separate.
3. Combine saturated n-butanol and glacial acetic acid in the ratio of 4:1as this can be used as mobile phase.

Procedure:

For ascending paper chromatography:

1. Cut the paper into a desired rectangular size
2. Draw a line with pencil at least 2 cm from the bottom
3. Take the little quantity of sample using capillary tube and load on the paper as spot A, B and C respectively.
4. Pour 10 ml of mobile phase in the beaker and keep a side for 15 min.
5. Now, keep the paper in beaker and allow it to develop with mobile phase.
6. Take out the paper and air dry it.
7. Spray ninhydrin reagent and determine R_f values.

For easy understanding, follow youtube reference videos available at

https://www.youtube.com/results?search_query=paper+chromatography

or
https://goo.gl/bddmHl

scan

Radial paper chromatography:

1. Cut the paper to form a circular disc.
2. Do small oval opening at the center of the disc.
3. Draw a circular line with pencil around the center.
4. Load the sample A, B and C on the circular line with equidistance.
5. Keep cotton wick through narrow opening.
6. Fill the petridish with mobile phase.
7. Keep the paper on the petriplate and allow it to develop under lid.
8. Take out the lid and air dry the paper.
9. Spray ninhydrin reagent and dry the paper not exceeding 100 degree C in hot air oven.
10. Determine R_f and unknown sample.

Report

The R_f values of sample A, B, C was found to be:

Note

Based on Rf values we can compare the presence of components in the mixture.

References:

1. http://www.chemguide.co.uk/analysis/chromatography/paper.html
2. http://wiki.answers.com/Q/Disadvantages_of_using_paper_chromatography
3. http://www.buzzle.com/articles/paper-chromatography.html
4. http://test.scoilnet.ie/Res/johndaly120899170605_2.htm
5. http://en.wikipedia.org/wiki/Ninhydrin

2. THIN LAYER CHROMATOGRAPHY

Dr. Manish Kumar Thimmaraju

Aim:

To isolate into individual amino acids from the given mixture of amino acids using thin layer chromatography.

Requirements:

Apparatus required:

TLC plates, Glass beakers, 10 ml measuring cylinder and capillary tubes etc.

Chemicals required:

n-butanol, glacial acetic acid, distilled water (4:1:5), amino acids (proline) and ninhydrin reagent.

Principle:

Thin-layer chromatography consists of two phases. One is stationary phase and the other is mobile phase. TLC is conducted on sheet of plastic, glass or aluminum coated with a thin layer of cellulose, silica gel or aluminum oxide. Adsorption plays an important role in resolution of two components.

Advantages:

1. Less development time.
2. Capability to resolve compounds easily.
3. Fast separation process
4. Selectivity for compound is higher.
5. Inexpensive technique.

Disadvantages:

1. Limited separation ability.
2. Poor detection limit.
3. Sensitive to humidity and moisture.

Applications:

TLC helps in identification of

1. Purity of drugs and active pharmaceutical ingredients.
2. Determination of acids, alcohols, proteins, alkaloids, amines, antibiotics.
3. To determine reaction intermediates.
4. To determine efficiency of resolution processes.
5. being a semi quantitative technique.

Solvent preparation:

1. Saturate n-butanol with water for 24 hours.
2. Separate above using funnel.
3. Combine saturated n-butanol and glacial acetic acid in the ratio of 4:1as this can be used as mobile phase.

Procedure:

1. As similar to paper chromatography, cut the pre-coated aluminum plates which can be freely immersed in the beaker.
2. Pour little quantity of mobile phase in the beaker and cover the lid.
3. Draw a baseline on the TLC plate using pencil.
4. Load the samples A, B and C on TLC plate an allow them to dry.
5. Keep this plate in TLC beaker.
6. Wait till solvent reaches 80% of TLC plate.
7. Take out the plate and spray ninhydrin reagent.
8. Determine R_f values.

Report

The R_f values of sample A, B, C was found to be:

Note

Based on Rf values we can compare the presence of components in the mixture.

For easy understanding, follow youtube reference videos available at

https://www.youtube.com/results?search_query=thin+layer+chromatography+experiment

or
https://goo.gl/6X7kS7

scan

References:

1. http://bheem.hubpages.com/hub/tlc-thin-layer-Chromatography-Principle-procedure
2. http://en.wikipedia.org/wiki/Ninhydrin
3. http://en.wikipedia.org/wiki/Thin-Layer_chromatography.
4. http://chemwiki.ucdavis.edu/VV_Lab_techniques/Thin-Layer_chromatography

3. THE SOLUBILITY OF ACTIVE PHARMACEUTICAL INGREDIENT

Dr. Manish Kumar Thimmaraju

Aim:

To determine the solubility of the given Active Pharmaceutical Ingredient in different solvent systems.

Apparatus Required:

10 ml Standard volumetric flasks, test tubes, test tube stand, pH meter, etc.

Chemicals Required:

Double distilled water, 20 % v/v methanol, 20% v/v ethanol, 0.1N HCl (it has pH 1.2), 1% w/v sodium lauryl sulfate, pH 6.8-phosphate buffer, pH 7.4-phosphate buffer, API(given unknown).

Principle:

If an unknown drug is given to our hand and allowed to analyze using either UV/Vis or HPLC or even flame photometry, our first duty is to determine the solubility of drug in one of the solvent systems. While selecting the various solvents, one should consider that the pH of the solvents should not be in the extreme range.

Our goal is to determine the solubility of any drug in the various solvents. Select various inorganic buffers to ensure that they do not have any absorbance in UV/Vis region.

If the drug is freely soluble in more than one type of solvents, we need to consider those solvents for further analysis.

Preparation of solvent systems:

1. 0.1 N Hydrochloric acid :

Dissolve 0.85ml of hydrochloric in 1000ml of distilled water.

2. Phosphate buffer p^H 6.8:

50ml of 0.2M potassium dihydrogen phosphate in 200ml standard volumetric flask and add 22.4ml of 0.2M sodium hydroxide and adjust the pH. Add distilled water to make up required volume.

3. Phosphate buffer p^H 7.4:

50ml of 0.2M potassium dihydrogen phosphate in 200ml standard volumetric flask and add 39.1ml ml of 0.2M sodium hydroxide and adjust pH. Add distilled water to make up required volume.

4. 1% w/v Sodium lauryl sulphate:

Dissolve 1gm of sodium lauryl sulphate in 100ml of distilled water in a 100ml standard volumetric flask.

5. 20% v/v Methanol:

Transfer 20ml of methanol to 100ml standard volumetric flask and make up to 100ml using distilled water.

**********Read Important Note after this page*********

6. 20% v/v Methanol in 0.1N hydrochloric acid:

Transfer 20ml of methanol to 100ml standard volumetric flask and make up to 100ml by using 0.1N hydrochloric acid.

7. 20% v/v Methanol in pH 6.8 phosphate buffer:

Transfer 20ml of methanol to 100ml standard volumetric flask and make up 100ml by using 2% Phosphate buffer pH 6.8.

8. 20% v/v Methanol in pH 7.4 phosphate buffer:

Transfer 20ml of methanol to 100ml standard volumetric flask and make up to 100ml by using 2% Phosphate buffer pH 7.4.

******Important Note******

Transfer accurately weighed quantity 10 mg of API directly to each solvent 100 ml mentioned between 1 and 5 and shake vigorously to obtain clear solution.

For those solvents mentioned between 6 and 8, take 10 mg of API, firstly dissolve in organic solvent and then make up to 100 ml with water.

Report:

1	0.1 N Hydrochloric acid	Insoluble
2	Phosphate buffer p^H 6.8	Insoluble
3	Phosphate buffer p^H 7.4	Insoluble
4	1% w/v Sodium lauryl sulphate	Insoluble
5	20% v/v Methanol	Soluble
6	20% v/v Methanol in 0.1N hydrochloric acid	Soluble
7	20% v/v Methanol in pH 6.8 phosphate buffer	Soluble
8	20% v/v Methanol in pH 7.4 phosphate buffer	Soluble
9	Water	Soluble

Reference:

1. Indian Pharmacopoeia 2007

4. DETERMINE THE MAX. ABSORBANCE OF POTASSIUM PERMANGANATE BY UV/VISIBLE SPECTROSCOPY

Dr. Manish Kumar Thimmaraju

Aim:

To find the λ max of potassium permanganate using UV/VISIBLE spectroscopy.

Apparatus Required:

UV/VISIBLE spectrophotometer, 10ml standard volumetric flask, micropipette, 100ml standard volumetric flask, potassium permanganate and distilled water.

Principle:

The equipment for ultraviolet-visible spectroscopy is called a UV/Vis spectrophotometer which is essential for determining the intensity of light passing through a drug sample (I), and compared it to the intensity of transmitted light (I_o). I/I_o is called the transmittance, measured as a percentage (%T).

The absorbance, A, is derived from transmittance:

$$A = -\log(\%T/100\%)$$

Applications:

For qualitative and quantitative determination of conjugated organic compounds, proteins, nucleic acids.

Procedure:

Preparation of stock solution:

Transfer 10mg of potassium permanganate to 100ml volumetric flask and make up to required volume with water.

Preparations of the samples:

µg/ml	Stock solution volume ml	Water volume ml
2	0.2	9.8
4	0.4	9.6
6	0.6	9.4
8	0.8	9.2
10	1	9

Determination of λ max by using uv-visible spectrometer:

- Transfer water to cuvette and take baseline measurements between 400-600nm.
- Find out the wavelength of maximum absorbance (λ max).

Results:

µg/ml	Wavelength	Absorbance
	524	0.022
10	525	0.089
	526	**0.147**
	527	0.122

Report:

The λ max of potassium permanganate was found to be 526 nm.

REFERENCES:

1. https://en.wikipedia.org/wiki/Ultraviolet–visible_spectroscopy#Ultraviolet-visible_spectrophotometer

5. MOLAR ABSOPTIVITY OF KMnO$_4$

Vandana Pamulaparthy

Aim:

To find the molar absoptivity of KMnO$_4$.

Requirements:

Micropipette, volumetric flask, potassium permanganate.

Principle:

(i) Molar absorptivity is the ratio of absorbance to the concentration in moles per liter multiplied by the path length in cm.

$$\text{Molar Absorptivity, } \varepsilon = A / c\,l$$

Where   A = absorbance,

c = sample concentration in moles/liter &

l = length of light path through the sample in cm

Procedure:

Prepare 10µg/ml of potassium permanganate and determine the absorbance using uv-vis spectrophotometer.

At 526nm the absorbance is 0.147.

The molecular weight of potassium permanganate is 158.034 g/mol.

10 micrograms of potassium permanganate = 63.2775 x 10^{-9} moles.

Molar Absorptivity, $\varepsilon = A / c\, l$

$A = 0.147$

C = should be in moles / liters

$10\mu g/ml$ can be said like => 1 ml contains 10 micrograms

1000 ml contains 10000 micrograms

So, 10 mg in 1 liter.

Convert 10 mg(0.01g) to moles.

The molecular weight of potassium permanganate is 158.034

$$158.034\, g \quad = \quad 1\, mole$$

$$0.01g \quad = ?$$

$$? = (0.01 \times 1)/158.034 =$$

$$(6.32 \times 10^{-5}) = C;$$

$$\varepsilon = 0.147/6.32 \times 10^{-5} = \mathbf{2325\ lit.\ mol^{-1}.\ cm^{-1}}$$

Since, L=1 cm

Report:

The molar absorptivity of the $KMnO_4$ was found to be **2325 lit. mol^{-1}. cm^{-1}**

REFERENCE:

1. Y.R.Sharma, 2010, Elementary Organic Spectroscopy.

6. DEVELOPING CALIBRATION GRAPH FOR GIVEN DRUG

Vandana Pamulaparthy

Aim:

To develop a calibration curve for a given drug.

Apparatus required:

Standard volumetric flasks (10ml and 100ml) and micropipette.

Principle:

Calibration curve which generates the relation between concentration and absorbance. The minimum and maximum absorbance for developing a calibration curve should lie between 0.1 and 1.0. If absorbance exceeds the value 1 then dilute the sample and proceed for further absorbance measurements till we should get absorbance value less than 1.

Stock solution preparation:

Transfer 10mg of the drug sample to 100ml volumetric flask and add 20% methanol with phosphate buffer pH 6.8 (we can select solvent as per previous experiment).

Preparation of various concentrations:

Required Concentration μg/ml	Required Stock solution (ml)	Required Solvent for dilution
6	0.6	9.4
8	0.8	9.2
10	1	9
15	1.5	8.5
20	2	8
30	3	7
40	4	6

Determine the absorbance for each concentration at particular wavelength.

Perform triplicates of concentrations.

Observation:

Concentration μg/ml	Absorbance at 276 nm
6	0.154
8	0.189
10	0.235
15	0.351
20	0.475
30	0.751

Calibration graph

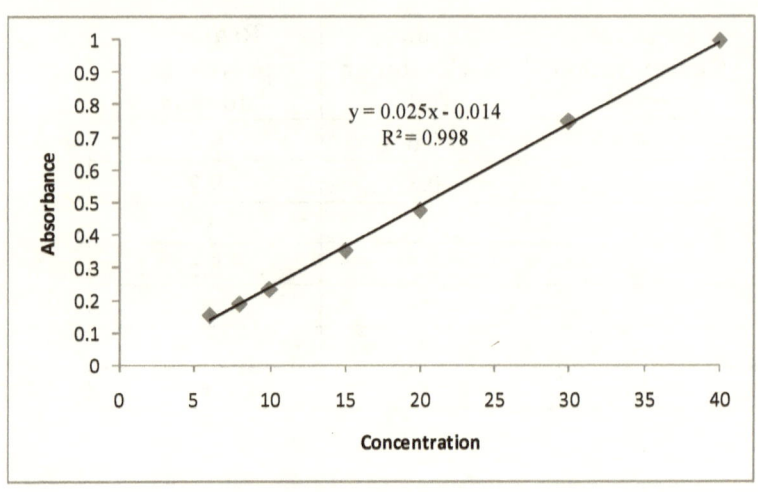

- The correlation coefficient was found to be R^2 is 0.998
- The linear equation is

Absorbance= 0.025 x Conc +0.014

Reference:

1. https://en.wikipedia.org/wiki/Calibration_curve
2. https://terpconnect.umd.edu/~toh/models/CalibrationCurve.html

7. FT-IR Spectroscopy

Dr. Manish Kumar Thimmaraju

Aim:

To determine the functional groups present in given unknown drug samples by using FT-IR.

Requirements:

Samples I and II, FT-IR, polystyrene, acetone, agate mortar and pestle.

Sample Preparation:

Pellet technique –

In this method, potassium bromide is pulverized with given drug and compacted to form disc in hydraulic press.

For easy understanding, follow youtube reference videos available at
https://www.youtube.com/results?search_query=kbr+pellet

or

https://goo.gl/Q4Cg6Y

scan

Procedure:

1. Transfer potassium bromide powder approx. 100mg and add small amount 5 mg of drug to it. Finely pulverize the mixture using agate mortar.

2. Transfer the powder to stainless steel die set and keep under hydraulic press.

3. Take out clear pellet sample and keep in FT IR instrument.

4. Perform analysis and determine the functional groups.

Factors for Cloudy Disks:

1. Potassium bromide is not powdered properly.

2. Wet samples.

3. Too thick pellets.

4. Powder is not distributed uniformly of steel disc.

Results:

Sample I

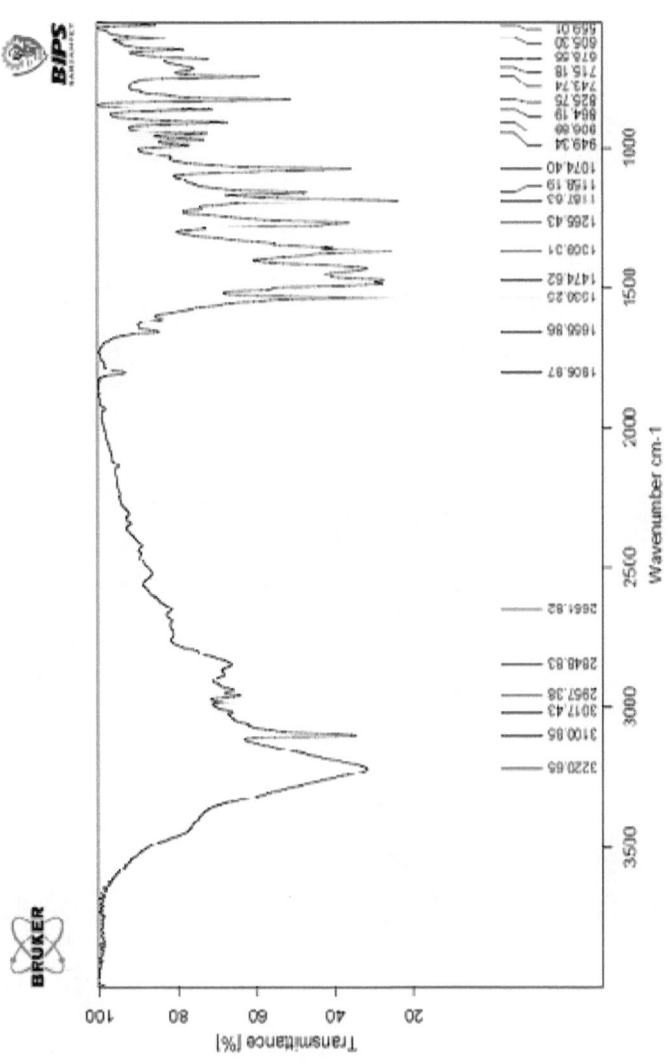

Sample II

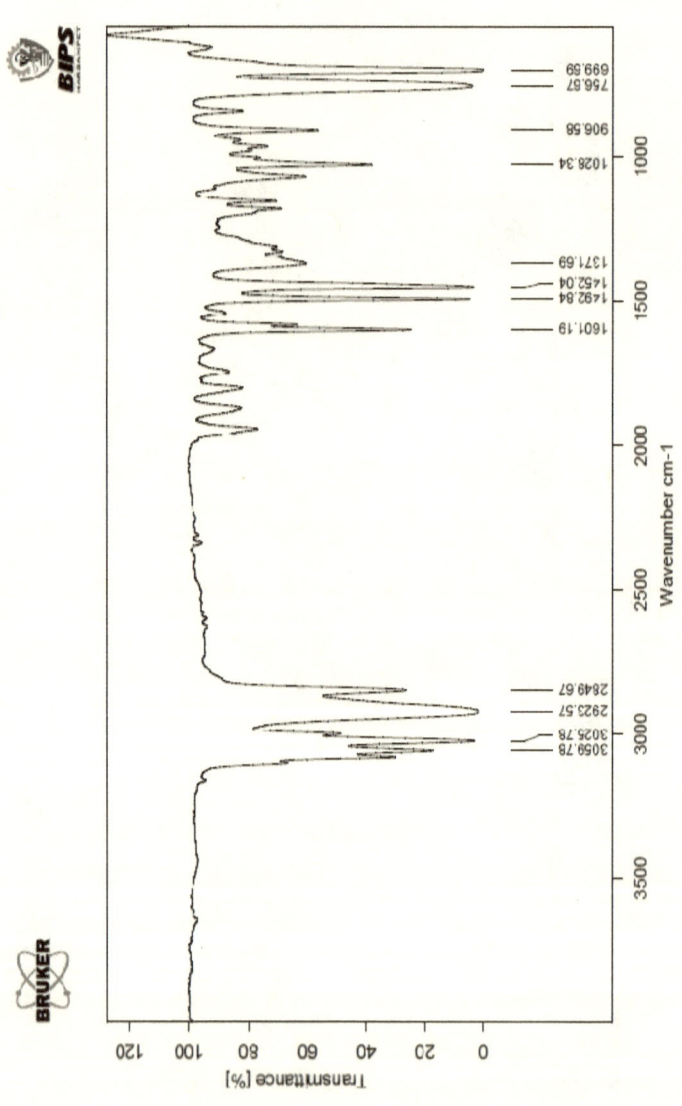

Wave numbers	Functional groups
3220	Carboxylic acids,amides
3100	Carboxylic acids,amides
1655	Carbonyls
1536	Nitro compounds
1474	Aromatics
1369	alkanes
1265	Carboxylic acids,amides
1187	Alcohols,esters,ketones,anhydrides
1158	Alcohols,esters,ketones,anhydrides
1074	anhydrides
949	anhydrides
906	benzenes
864	alcohols
825	benzenes
743	acid chlorides
605	acid chlorides

Wave numbers	Functional groups
3059	Carboxylic acids
1601	Carbonyls
1492	Aromatics
1371	Alkanes
1026	Alcohols,anhydrides
906	anhydrides
699	acid chlorides

8. DETERMINATION OF VISCOSITY OF VARIOUS POLYMERIC SOLUTIONS

Dr. Manish Kumar Thimmaraju

Aim:

To determine the viscosity of various polymeric solutions using Brook field viscometer.

Requirements:

Brook field viscometer DV-II Pro, polymeric solutions (HPMC, Acacia), volumetric flasks.

Principle:

Viscosity is the resistance to flow and always depends on temperature and pressure.

A shear stress acts across the surface, parallel to the surface.

Rate of shear is a gradient of velocity in a flowing material.

Procedure:

Prepare different polymeric solutions such as1% HPMC, 1% acacia.

Preparation of different of polymeric solutions:

1% w/v HPMC:

Transfer 1 gram of HPMC to standard volumetric flask, dissolve in little quantity of water and make up to 100ml with water.

1% w/v Acacia:

Transfer 1 gram of acacia to standard volumetric flask, dissolve in little quantity of water and make up to 100ml with water.

Measurements of various parameters using 1% HPMC :

RPM	shear rate	shear stress	viscosity	torque
10	12.23	0.04	0.42	0.6
25	30.58	0.54	1.78	7.4
70	85.61	2.11	2.47	28.8
100	122.3	3.18	2.6	43.4
150	183.5	5	2.72	68.1
196	239.7	6.72	**2.8**	**91.6**

1 % w/v HPMC has viscosity 2.8 cps

(Viscosity at highest torque is considered)

Measurements of various parameters using 1% acacia :

RPM	shear rate	shear stress	viscosity	torque
10	12.2	0.14	1.14	1.9
25	30.53	0.34	1.1	4.6
70	85.61	0.95	1.11	12.9
100	122.3	1.69	1.39	23
150	183.5	3.29	1.81	45
196	239.7	5.03	**2.12**	**69.4**

1 % w/v Acacis has viscosity 2.12 cps

(Viscosity at highest torque is considered)

For easy understanding, follow youtube reference videos available at

https://www.youtube.com/results?search_query=bookfield+dv+pro

or

https://goo.gl/WBwhFI

scan

Reference:

http://www.mne.psu.edu/cimbala/Learning/Fluid/Introductory/what_is_fluid_mechanics.htm

9. DISSOLUTION METHOD FOR A GIVEN TABLETS USING USP-II APPARATUS

Vandana Pamulaparthy

Aim:

To find the percentage drug release of given tablets by dissolution.

Requirements:

Tablets, phosphate buffer pH 6.8, test tubes, 10ml pipette

Principle :

It is essential for determining the batch to batch uniformity of drug release and to determine bioequivalence.

There are Seven types of U.S.P apparatus are specified...

1. Apparatus 1 (basket apparatus)
2. Apparatus 2 (paddle apparatus)
3. Apparatus 3 (reciprocating cylinder)
4. Apparatus 4 (flow through cell)

Paddle apparatus:

It has vessel having the capacity of 1 liter and usually 900 ml is filled with dissolution media. It consists of paddle and operated with a variable speed motor. This apparatus is used for quality testing of uncoated tablets, sublingual tablets, hard gelatin capsules and soft gelatin capsules, enteric coated tablets, gels, ointments.

Procedure for development of calibration curve for pure API:

Transfer 10 mg of drug to 100ml capacity standard volumetric flask and add solvent to solubilize the drug. Then prepare various calibration concentrations like 10, 15, 20, 40, 60µg/ml and observe the absorbance.

Conc.	Abs-I	Abs-II	Abs-III	Mean Abs.
10	0.152	0.149	0.151	0.15067
15	0.221	0.219	0.223	0.221
20	0.305	0.301	0.303	0.303
40	0.608	0.603	0.612	0.60767
60	0.921	0.929	0.924	0.92467

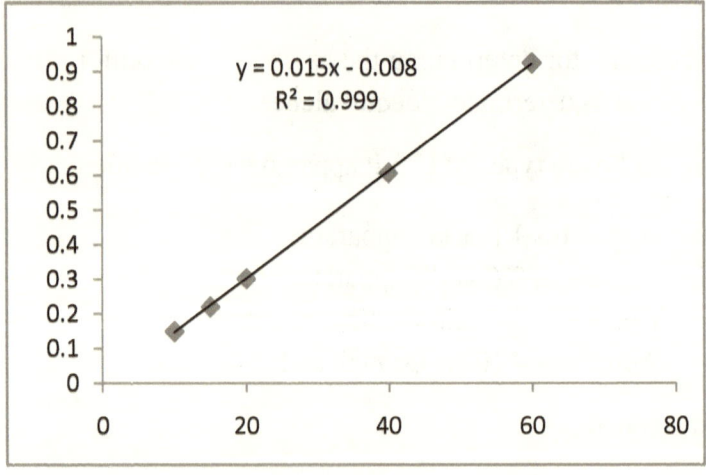

$y = 0.015x - 0.008$

$R^2 = 0.999$

X axis- concentration

Y axis Absorbance

Preparation of dissolution medium for tablets:

Dissolve 13.6gms of potassium dihydrogen phosphate and adjust the pH 7.4 and dilute to1000ml water.

Dissolution Procedure:

1. Fill the dissolution chamber with 900ml of buffer.
2. Place the tablets in the dissolution medium.
3. Adjust the temp. to 37 degree C.
4. Fix rpm to 50 or as per pharmacopoeia.
5. Withdraw 10ml of sample at 30, 60, 90,120,150,180 and 360 minutes.
6. Replace the 10ml sample with fresh buffer to maintain sink conditions.
7. Analyze the sample using UV and determine absorbance.

Time	Abs1	Abs2	Average	Conc(mg)	% Cumulative Drug Release
0	0	0	0	0	0
30	0.221	0.211	0.216	13.44	53.76
60	0.298	0.282	0.29	17.88	71.52
90	0.313	0.278	0.2955	18.21	72.84
120	0.284	0.271	0.2775	17.13	68.52
180	0.314	0.317	0.3155	19.41	77.64
360	0.399	0.415	0.407	24.9	99.6

Report:

The amount of drug release from the formulation at 360 min is 99.60%.

Reference:

1. https://en.wikipedia.org/wiki/Dissolution_testing

10. COMPARISON OF DISSOLUTION PROFILES OF TWO DRUGS

Vandana Pamulaparthy

Aim:

To determine the difference between the dissolution profiles of two different marketed formulations using f_2 method.

Requirements:

Tablet formulations, phosphate buffer pH 7.4, test tubes, 1ml micropipette.

Principle:

Similarity factor (f_2):

f2 method is the simplest of the all methods mentioned in the literature to compare the dissolution profiles of pharmaceutical dosage forms.

f_2 measures the proximity between the two formulation dissolution profiles. f_1 is known as difference factor, and f_2 is known as similarity factor.

$f_2=100$, if two profiles are similar. The value of f2 should be always greater than 50.

$$f_2 = 50 \log\{[1 + \frac{1}{n}\sum_{t=1}^{n} (R_t - T_t)^2]^{-0.5} \times 100 \}$$

n = number of time points.
Rt = cumulative % drug release at 'n' time points of the reference product.
Tt = cumulative % drug release at 'n' time points of the test product.

Reference cumulative %drug release (R)	Test cumulative %drug release (T)
8.07	12.87
18.78	29.94
31.53	47.16
46.74	63.54
65.16	82.23
89.82	106.89

Reference cumulative %drug release	Test cumulative %drug release	R-T
8.01	12.86	-4.85
18.79	37.85	-19.06
31.57	47.24	-15.67
46.77	63.78	-17.01
63.5	82.89	-19.39
88.25	95.86	-7.61

(R-T)2	1/n	1/n (Sigma)	
23.5225	0.166666667	225.929883	
363.2836			f2
245.5489	1+1/n(sigma)	Power -0.5	41.10271
289.3401	226.9298833	0.06638258	
375.9721			
57.9121	x 100	6.63825842	
1355.58			

31

Report:

The f_2 value was found to be 41.10 indicates difference between Reference and Test.

Reference:

1. http://www.dissolutiontech.com/DTresour/899Art/DissProfile.html

11. DETERMINATION OF UNKNOWN CONCENTRATION OF CAFFEINE USING HPLC

Vandana Pamulaparthy

Aim:

To determine the unknown concentration of caffeine using isocratic HPLC

Requirements:

Standard volumetric flasks, HPLC grade methanol, caffeine, triple distilled water.

Principle:

In this isocratic HPLC, the composition of mobile phase constant throughout the analysis. The composition of mobile phase for the determination of caffeine is 70% v/v methanol. HPLC grade methanol is transferred to Millipore water and further sonicated to remove dissolved gases. Furthermore, Hamilton syringe is used to inject the samples. All the calibrations concentrations can be made of 70% methanol. The wavelength of detection was 270 nm.

Procedure:

Transfer 10 mg of caffeine to 100ml standard volumetric flask and add 70 ml of HPLC grade methanol. Solubilize caffeine and add

Millipore water to make up final volume. Prepare three calibration concentrations such as 5 μg/ml , 10μg/ml and 15 μg/ml using above stock solution.

5 μg/ml solution is prepared by withdrawing 0.5 ml from standard stock solution and made up to 10 ml with 70% v/v methanol. Similarly 10 μg/ml concentration is prepared by withdrawing 1 ml from standard stock and dilute to 10 ml with 70% v/v methanol. In the same way, 15 μg/ml was prepared. Inject the sample in to HPLC and determine the area associated with each concentration.

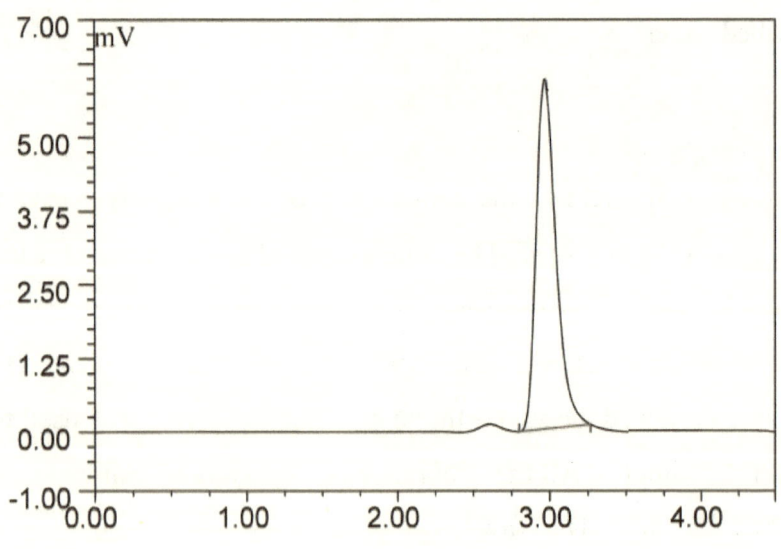

X axis-Time

Sample chromatogram of caffeine 5 μg/ml at 270 nm

Concentration μg/ml	Area
5	311.26
10	459.75
15	614.49
Unknown	512.1

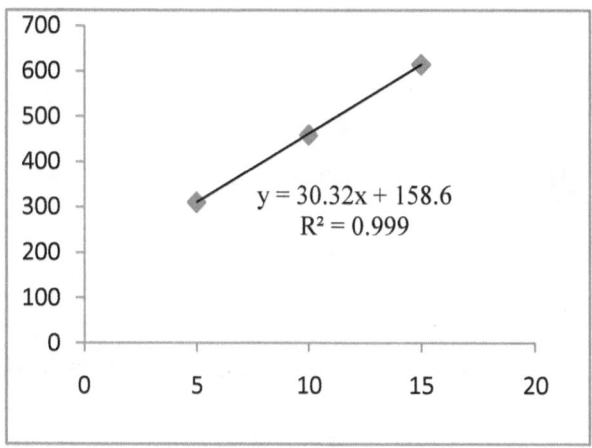

$y = 30.32x + 158.6$
$R^2 = 0.999$

X axis-concentration
Y-axis-peak area

Unknown concentration calculation:

$$(512.1 - 158.6) / 30.32 = 11.65 \ \mu g/ml$$

Report: The unknown concentration of caffeine was found to be

11.65 μg/ml

12. CALIBRATION GRAPH FOR ONDANSETRON USING UV SPECTROSCOPY

Vandana Pamulaparthy

Aim:

To develop calibration graph for ondansetron using UV spectroscopy

Requirements:

Ondansetron API, standard volumetric flasks, 1 ml micropipette, methanol.

Procedure:

Preparation of stock solution and various concentrations:

Transfer accurately weighed quantity 10 mg of ondansetron to 100 ml standard volumetric flask and add 40 ml of methanol. Further dilute the solution to 100 ml. Withdraw 0.2 ml from the above solution and make up to 10 ml with 40% methanol to obtain 2 μg/ml concentration. Similarly, prepare 4 μg/ml, 6 μg/ml, 8 μg/ml, 10 μg/ml, 12 μg/ml and 15 μg/ml and determine the absorbance.

Concentration (μg/ml)	Mean Abs.
2	0.216
4	0.331
6	0.403
8	0.523
10	0.644
12	0.742
15	0.921
Unknown	0.487

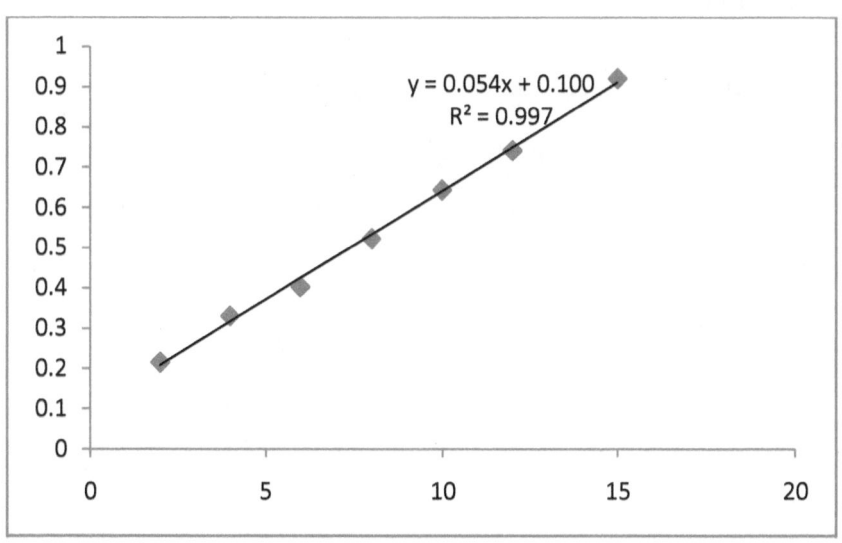

$y = 0.054x + 0.100$
$R^2 = 0.997$

**X axis- concentration
Y axis Absorbance**

Unknown absorbance was is 0.487 then,

$$(0.487 - 0.1) / 0.054 = 7.16 \ \mu g/ml.$$

Report:

- The calibration curve was developed between 2 μg/ml to 15 μg/ml.

- The co-relation co-efficient was found to be $R^2 = 0.997$.

References:

1. http://en.wikipedia.org/wiki/Calibration_curve

13. CALIBRATION GRAPH FOR ACECLOFENAC USING UV SPECTROSCOPY

Vandana Pamulaparthy

Aim:

To develop calibration graph for aceclofenac using UV spectroscopy

Requirements:

Aceclofenac API, standard volumetric flasks, 1 ml micropipette, methanol.

Procedure:

Preparation of stock solution and various concentrations:

Transfer accurately weighed quantity 10 mg of aceclofenac to 100 ml standard volumetric flask and add 30 ml of methanol. Further dilute the solution to 100 ml. Withdraw 0.2 ml from the above solution and make up to 10 ml with 40% methanol to obtain 2 µg/ml concentration. Similarly, prepare 4 µg/ml, 6 µg/ml, 8 µg/ml, 10 µg/ml, 12 µg/ml and 15 µg/ml and determine the absorbance.

Concentration (µg/ml)	Mean Abs.
4	0.121
6	0.249
8	0.349
10	0.451
12	0.579
15	0.682
20	0.872
Unnknown	0.651

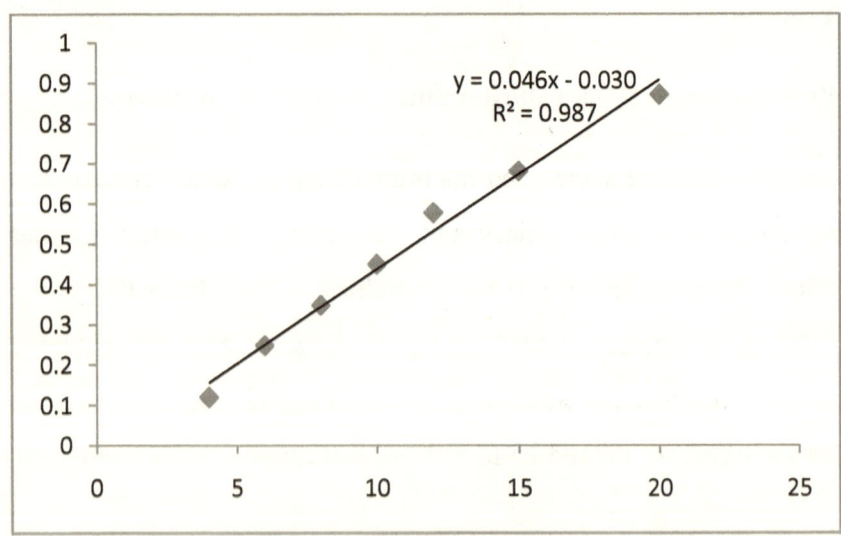

y = 0.046x - 0.030
R^2 = 0.987

X axis- concentration
Y axis Absorbance

Unknown absorbance was is 0.651 then,

$$(0.651 + 0.03) / 0.046 = 14.8 \ \mu g/ml.$$

Report:

- The calibration curve was developed between 4 μg/ml to 20 μg/ml.

- The co-relation co-efficient was found to be $R^2 = 0.987$.

References:

1. http://en.wikipedia.org/wiki/Calibration_curve

14. IPQC TESTS FOR FORMULATIONS

Vandana Pamulaparthy

Aim:

To formulate the tablets and perform in-process quality control tests on immediate release tablets.

Chemicals required:

Paracetamol, talc, stearic acid, microcrystalline cellulose, starch paste, flavouring agent, colouring agent, sweetners and starch.

Instruments required:

Friabillator, monsanto hardness tester, disintegration test apparatus, screw guage or vernier caliper, tablet compression machine, motor and pestle,

Principle:

The IPQC tests for tablets should undergo:

1. Weight variation test

2. Thickness test

3. Hardness test

4. Friability test

5. Disintegration test

Formulation of paracetamol (300mg) tablet

Paracetamol	250mg
Mirocrystalline cellulose	125mg
Starch	65 mg
PVP	5% solution
Talc	1.5mg
Stearic acid (lubricant)	10mg

Procedure:

Preparation of Immediate Released Tablets:

1. Prepare 5% PVP in 100ml of water.
2. Add slurry to remaining excipients except stearic acid.
3. Sieve the above mass through no-22 screen and keep in hot air oven.
4. Further compress dried granules into tablets.

Weight variation test

Formulation	Mean weight	% deviation
Uncoated and film coated tablets	80 mg or less	10
	More than 80 mg but less than 250 mg	7.5
	250 mg or more	5

Thickness test:

The tablet thickness is expected to be within ±5% of the prescribed values and measured with calipers.

Hardness test:

The hardness of tablet should lie in between 3kg and 7kg.

Friability test:

Friability is the resistance to shock and abrasion during manufacture, packing, convey and while using. Roche friabilator consists of transparent drum which has arm shaped blade and can be rotated with a speed of 25rpm. A loss of not greater than 1.0 per cent is acceptable for most of the tablets.

Disintegration test:

All the tablets taken for the DT should pass the test. If 1 or 2 tablets fail then proceed with 12 additional tablets. 16 tablets out of 18 should disintegrate.

Weight variation test:

$$\frac{\text{Indiviual tablet weight} - \text{Average weight}}{\text{Average weight}} \; x \; 100 = \% \text{ weight variation}$$

Thickness test :

$$\frac{\text{Indiviual tablet thickness} - \text{Average thickness}}{\text{Average thickness}} \; x \; 100$$

Friability test:

Initial weight of 20 tablets = w_1

Final weight of 20 tablets = w_2

Weight difference   = $w_1 - w_2$

$$\text{Friability} = \frac{w1 - w2}{w1} \; x \; \mathbf{100}$$

Report:

IPQC Test		Practical Value	IP limit
Weight variation	Min= Max =		100±5 %
Thickness	Min = Max =		100±5%
Friability			Not more than 1%
Hardness	Min = Max =		3-7 kg/cm^2
Disintegration			30 min

References:

1. Pharmaceutical Methods, Indian Pharmacopeia, volume-I, 2007, Page.No 175-183
2. http://shodhganga.inflibnet.ac.in/bitstream/10603/8541/18/18_chapter%206.pdf

15. SYSTEM SUITABILITY PARAMETERS

Vandana Pamulaparthy

Aim:

To calculate system suitability parameters for HPLC.

Requirements:

Millipore water, acetonitrile, rosuvastatin calcium, HPLC instrument, standard volumetric flasks, thermo scientific, finnipipette, vacuum filtration apparatus.

Principle:

Various System suitability parameters are:

Adjusted retention time:

The tenure of drug molecule in the stationary phase and is given by

$$t^1_R = t_R\text{-}t_M \text{ sec.}$$

The value t_M is obtained an unretained component. Eg: air or methane.

Retention time:

It is the sum of the times of drug present in the mobile phase (t_M) and in the stationary phase (t^1_R).

$$t_R = t^1_R + t_M$$

Retention volume:

This is the volume of mobile phase required to elute one half of the drug substance from the column.

$$V_R = t_R \times f$$

f = flow rate of the mobile phase.

Capacity factor (k):

It is the ratio of the time the drug present in the stationary phase to the time in the mobile phase and capacity factor should be more than 2.

$$k = t^1_R / t_M = t_R - t_M / t_M$$

No.of theoretical plates:

A theoretical plate is an equilibrium between stationary phase and mobile phase. It does represent the efficiency of a column and for a given column, the value should be 2,000.

$$N = \frac{L}{H} \text{ or}$$

$$N = 16 \, (t_R/w)^2$$

HETP (height equivalent to a theoretical plate):

It is the distance travelled by a solute during partitioning and important in the efficiency of separation and the value should be less.

$$H = \frac{L}{N}$$

Symmetry factor:

Symmetry is calculated using = (A+B)/2B

A = the distance from peak centre to right tailing end

B = the distance from peak centre to left fronting edge

The value should be should be less than 2.

Velocity of mobile phase:

$$u = L / t_M \text{ cm/sec} ;$$

$$L = \text{length of column}$$

Velocity of drug molecules:

$$\tilde{v} = L / t_R \text{ cm/sec}$$

Resolution:

It is the ratio of distance between band peaks to average bandwidth and the value should be more than 2.

$$\mathbf{R_S} = 2\ [(t_R)_B - (t_R)_A] / [W_A + W_B]$$

Width of the peak:

$$w = 4t_R/\sqrt{N}$$

Reference:

1. Douglas A.Skoog, F.James Holler, Stanley R.Crouch, An Introduction To Chromatographic Separations, Instrumentation Analysis, 2011, page no: 841-853.
2. A.H.Beckett, J.B.Stenlake-2004, 4th edition-part two, Chromatography, Practical Pharmaceutical Chemistry, page no: 138-143.

16. ANALYTICAL METHOD DEVELOPMENT FOR A GIVEN DRUG USING HPLC

Vandana Pamulaparthy

Aim:

To perform the AMD for a given drug using shimadzu gradient HPLC.

Requirements:

Standard volumetric flasks (10 ml and 100ml), micropipette, vacuum filtration apparatus, rosuvastatin calcium (10mg), millipore water, acetonitrile.

Principle:

AMD is essential for determining potency, impurities and degradation products during stability testing.

AMD using HPLC consists of following steps.

Preparation of triple distilled water
Water further processed to obtain millipore water
Determining the drug solubility in acetonitrile-water systems
Preparing various concentrations
Degassing mobile phase using bath sonicator
Fixing the λ max in the detector
Detector stabilization
Injecting various concentrations
Analyzing the response
Preparing the calibration graph –conc vs. area

Analysis parameters

Mobile phase	:	70% v/v acetonitrile
λ max	:	243nm
Flow rate	:	1ml/min
Pressure	:	440 kgf (max)
Detector	:	uv

#	Conc. (ŋg/ml)	Peak Area (mV.s)
1	100	13674
2	200	15982
3	500	24969
4	1000	39196
5	5000	146456
6	10000	293424
7	15000	394879

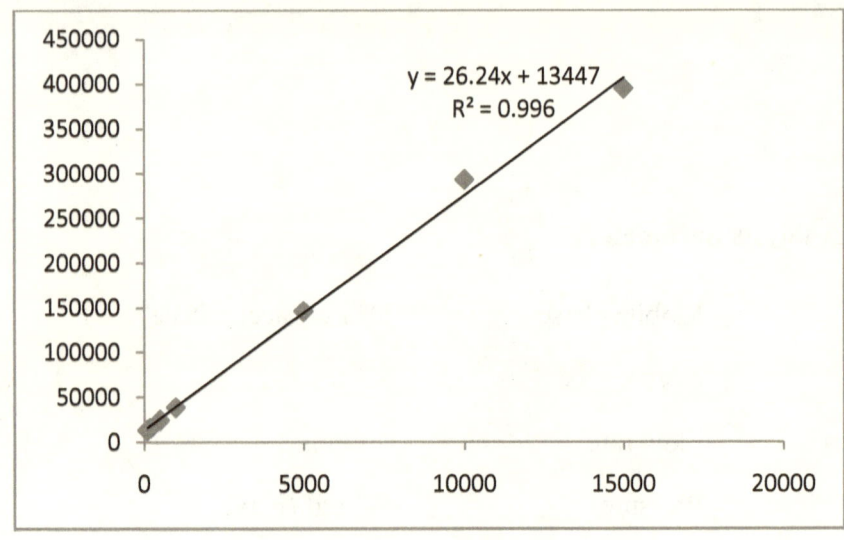

$y = 26.24x + 13447$
$R^2 = 0.996$

X axis-concentration
Y-axis-peak area

Report:

The analytical method was developed with a range 100 ng/ml to 15,000 ng/ml and found be linear with conc. Vs area.

Reference:

1. http://www.sciencedirect.com/science/article/pii/S0021967302015364

www.ingramcontent.com/pod-product-compliance
Lightning Source LLC
Chambersburg PA
CBHW021415170526
45164CB00002B/654